THE SPACE BEFORE YOUR FACE

by

Charles Walton

AuthorHouse™
1663 Liberty Drive, Suite 200
Bloomington, IN 47403
www.authorhouse.com
Phone: 1-800-839-8640

©2007 Charles Walton. All rights reserved.

No part of this book may be reproduced, stored in a retrieval system, or transmitted by any means without the written permission of the author.

First published by AuthorHouse 8/16/2007

ISBN: 978-1-4208-1112-4 (sc)

Printed in the United States of America
Bloomington, Indiana

This book is printed on acid-free paper.

1663 LIBERTY DRIVE
BLOOMINGTON, INDIANA 47403
(800) 839-8640
www.authorhouse.com

ACKNOWLEDGMENTS

-- The Los Gatos Unitarian Fellowship whose members share weekly in concert with their children the wonders of human life and the world around us, and:

-- My wife Ann Katherine who has encouraged me to get this work completed.

DEDICATED TO

-- Scientific American magazine for reporting on what is happening around us, and:

-- Richard Feynman, who spoke of the complex activity in the space before us; and:.

-- all the thinkers who came before us and shows us what was going on in space:

-- and to all the thinkers who follow and who will show us more…

CONTENTS

Your Cubic Foot 1

Molecules Of Gas Occupy The Space 2

Sound Goes Through The Cube 4

Wind Is Moving Air Molecules 6

Gravity Fields 8

Magnetic Field Of The Earth 10

Electric Field Of The Earth 12

Light Rays Pass Through 14

Radio Waves Pass Through 16

Short Radio Waves Pass Through 18

Location Signals From Space 20

Heat Waves Pass Through 22

X-Ray Radiation And Cosmic Rays 24

Neutrinos Pass Through 26

Flavored Molecules 28

Particles And Life In The Space Before Us 30

The Universe Speeds Through Our Space 32

Things Not Found In Our Space 34

Extra Sensory Perception 36

Mind Touches Mind 38

A Look Back 40

Measuring The Speed Of Light 43

Glossary 53

Appendix 55

Apparatus List 57

YOUR CUBIC FOOT

    Everyone has their own cubic foot of space before their face. What does it contain? *I will introduce you to wonders about that cubic foot which you will never forget, and it is all true.*

    What is in your cubic foot? What is going in one side and coming out the other? How do we tell what is going on in the cube? We need observation and we need sensors. Even after we look, and I tell you what we already know, there is yet more waiting for you and others to find. The operating principles of the universe, and more, are reflected or sensible in that cubic foot.

MOLECULES OF GAS OCCUPY THE SPACE

Molecules of oxygen, nitrogen, CO2, H20, and other gases fill the space. Air is made up of molecules or particles of nitrogen (80%), oxygen (18%), and other such as CO2 (carbon dioxide), H20 (water), and argon. The molecules are made of atoms. Human beings need air, especially oxygen, to keep their muscles working. We breath in oxygen and then we expel waste products such as carbon dioxide.

More on Molecules of Gas

The number of molecules of gas in a cubic foot of air is approximately 4.7 times one hundred thousand billion billion. Avogadro's number says there are 6.023×10^{23} molecules of gas per a weight of gas equal to the atomic number in grams. Oxygen is 16, nitrogen is 14; the average of air is close to 14. Molecules fly in and out of your cubic foot constantly, but the number going in is almost exactly the same as the number going out. One way to sense the presence of air is with a pressure gauge. The picture shows vanes suspended by a low friction bearing, such as needles, above and below the vaned device—this is a radiometer. The radiometer vanes will rotate inside a jar when radiation or light or heat strike the vanes. The vanes are black on one side and white on the other. When struck by light or heat the black side absorbs heat and warms up more than the white side. The black side becomes hotter than the white side. The molecules on the black side move more quickly than do those on the white side. The higher speed molecules develop more pressure than do the lower speed molecules. The result is motion of the vanes away from the black side. The vanes rotate away from the heat source. The speed of rotation is greater when the light source is greater."

Moisture is carried in the air, mostly invisibly. When the moisture content goes above a certain threshold, or the temperature falls below a threshold, the moisture "condenses out" as either drops of water or as a cloud.

MOLECULES OF GAS OCCUPY THE SPACE

If you guess that "nothing" is in the cubic foot, you are partly right. There is a lot of nothing between the molecules of air. There is a lot of molecules—actually there are seven hundred thousand billion billion molecules in a cubic foot of air.

SOUND GOES THROUGH THE CUBE

Sound passes through the space before us in the form of air waves. The waves are alternating compressions and rarefactions of the molecules of air. The sound waves of speech are produced by our tongue and throat, and the waves are sensed by an ear drum which couples to nerves within our ears. The nerves send a signal to the brain for analysis.

The speed of sound is approximately 1100 feet per second. Since a mile is 5,280 feet, this means that approximately five seconds are needed for sound to travel one mile. The speed of light is much faster. So—if you see lightning, count the number of seconds before you hear the thunder. It takes five seconds for sound to go one mile. "The time of light for one mile is 5 micro-seconds, or much less, or one millionth the sound time and is considered zero for this experiment." Divide the seconds by five to get the miles away the storm or lightning is. Measure twice or three times and you can tell whether the weather is getting closer or further away.

More About Sound Goes Through The Cube

Sound waves pass in all directions—only a limited number reach somebody's ears. Most sound is cyclical—that is there are repeated similar vibrations over a period of time. Music sound is produced by vibrating strings or membranes or columns of air. The vibration gets weaker or gets restarted, and changes frequency. In music some tone changes are beautiful and some are irritating. Sound can have a great range of frequencies, measured in octaves, where one octave is a doubling of the frequency. Sound varies in intensity, from a whisper to a shout to an explosion. The intensity is measured in decibels.

Your ear is just one type of detector. Another detector is a microphone, which has an electrical output. This output can be amplified and sent out on the air with much greater volume than the original. The electrical output signal can be recorded on magnetic tape or on glass surface discs (CDs), for playing at any time much later. Expert singers can make the same note, that is, the same frequency, sound joyful, or loving, or sad, or hateful—that is, emotion is conveyed by changing the qualities of the the note

SOUND GOES THROUGH THE CUBE

Our ears detect it. The air molecules vibrate and move our ear drums, which in turn stimulates the nerves connected to our brains.

WIND IS MOVING AIR MOLECULES

Molecules can be set in motion all in the same direction if there is some pressure difference between the two sides of the cube. Blowing air from our lungs into one side will create a pressure imbalance. We can sense or detect this molecular motion by dropping a piece of paper in the top. The blown air will push the paper out of the opposite side. If you place your hand in a moving stream of air, such as in an automobile with the window open, you can feel the pressure on your hand (Careful—don't put your hand outside.) If you hold a piece of string near the open window of the moving car, air flow will push the string almost horizontal, pointing to the back of the car. Moving molecules of air will push a falling leaf sideways if the wind is blowing. Rain and snow come down sideways when the wind blows.

More About Wind Is Moving Air Molecules

The air molecules have weight and inertia, and resist fast change in motion. The molecules cannot change speed without force being applied. When the sloping wing of an airplane moves through the air, the air is pushed down, and the air pushes back, or upwards, against the wing. Consequently, the airplane stays in the air... Wind is produced by the heat of the sun, combined with the presence of land and the rotation of the earth. At the shore, the sun heats the land during the day, more than the cool sea, the hot air over the land rises, and wind blows inshore. In the morning the land is cool, the ocean is warmer, and the wind blows offshore, or not at all. Out in the open ocean, the rotation of the earth plays an important part, and in the northern hemisphere, usually the wind blows West to East. To win a sailboat race, know your winds! Sailors often have pieces of string on the sails, to observe what the wind in their face is doing.

WIND IS MOVING AIR MOLECULES

Air molecules moving together make wind. We sense wine with a piece of blown paper, or by the waving of the trees, or by the feeling on our cheeks and arms, or viewing our breath on a cold day.

GRAVITY FIELDS

All objects in the universe have a gravitational attraction for all other objects. The earth beneath our feet is large and pulls things such as ourselves toward its center, with a force equal to our weight. The force goes through the cube before us. We can detect this force by letting an object fall through the cube in front of us, or just by jumping into the air. We always come down. Gravity holds the atmosphere of the earth around the earth.. Gravity holds the water of the rivers within their banks, and holds the ocean within its shores, and it holds the water within our drinking glass. If we invert the glass, gravity dumps the water..

More About Gravity Fields

The moon weighs one sixth as much as the earth and has one sixth the force of gravity, so that on the moon people weigh one sixth as much and can jump six times as high. Astronauts can jump high even with big back packs. A big object like Jupiter or the sun pulls harder on every thing; and a small object like the moon or a glass of water pulls less. The earth weighs 5.9×10^{27} grams (6.5 billion trillion tons) so it takes a lot of matter to make us feel heavy.

The moon pulls on the oceans of the world, and as it rotates around us, the ocean water moves, and tides are created. In the cube of space before our eyes. a dish of water inside the cube has "tides" in it. The tides are millionths of an inch high. If you study these little tides you can tell where the moon is, without using a telescope or looking at the sky. There is actually a rise of water towards the moon, and another away from the moon, owing to the gravitational gradient across the bowl. The sun is bigger but further away and produces yet smaller tides, so it is also possible to tell indirectly where the sun is.

Gravity is one of the four basic forces in the universe. It is considerably weaker than electric and magnetic forces or the nuclear forces, but we notice it because the masses of the earth and moon are so large. For typical matter the electric and magnetic fields are not noticed because typical clumps of matter have an equal umber of electric fields from electrons as positive fields from protons, and equal number of north and south magnetic poles. A black hole has such a strong gravitational field that not even light can escape. Gravitational fields exist between galaxies over zillions of miles and affect the rate at which the universe expands.

GRAVITY FIELDS

There is a strong gravity field from the earth. We sense this with a ball, which falls from the upper face to lower face of the cube. There are also gravity fields from the moon, the sun, and the stars.

MAGNETIC FIELD OF THE EARTH

The earth has a core formed largely of magnetic iron. The core forms a large magnet and creates a steady magnetic field flowing from pole to pole. The field flows through the space before our eyes, sideways mostly. The instrument which senses this field is a magnetized needle, free to move on a low friction bearing. The magnetic needle in this field points in the direction of the field. This pointing is very useful when sailing a ship around the world—the sailor can always tell which direction is North, south, East and West are.

More About The Magnetic Field Of The Earth

The magnetic poles of the earth are not in quite the same position as the rotational poles of the earth, so that if a mariner uses the compass to navigate, adjustments must be made for this mis-alignment. Also, at high latitudes, near the poles, the primary field dives downwards into the earth, and the needle position becomes weak and unstable.

The earth's magnetic field is generated in part by the iron core of the earth. The magnetic field is, overall, a loop of continuous magnetic "lines," which pass through the core, over the surface of the earth, and back to the iron core. The lines are generally all in a reliable pole to pole direction, but local iron deposits, such as in a mountain, or a buried meteorite, or a ship's engine, can distort the direction of the lines.

Over long periods of time it is known that the earth's magnetic field has varied sharply and reversed its polarity. This reversal leaves a trace in the rocks and is used to determine the age of other rocks.

The earth's magnetic field also plays a part in the aurora borealis, because the magnetic field affects the trajectories of particles radiated from the sun. The magnetic field creates the Van Allen belts. The magnetic field together with the ionosphere also protects life on earth from harmful radiation from the sun.

MAGNETIC FIELD OF THE EARTH

The earth's magnetic field reaches from the North pole to the South pole. We detect the earth's magnetic field with a compass. The compass needle rotates until it automatically lines up with the earth's field. The needle then points North and South.

ELECTRIC FIELD OF THE EARTH

The Earth's electric field reaches from the earth to the sky. It varies constantly, in part with the weather, and in large part with what conducting bodies are in the field. Our body is a weak conductor and disturbs the field. The field has a high voltage but the electrical resistance of the air is high and ordinarily there is such low current that we do not notice it. The field is a static or Direct Current ("DC") field, meaning that regular frequency Alternating Current (AC") fluctuations are not present. It may typically be 300 volts from the ground to the top of our heads.

More About The Electrical Field Of The Earth

The earth's electric field contributes to making lightning. We walk around and a voltage field follows us. The field can be detected electrically with high impedance electronic circuits, and then used to turn lights on and off, such as in a hand controlled lamp. Conductors rising into the sky, such as in a building equipped with lightning rods, or a steel frame, reduce the local field by bringing ground potential to the upper projection. Ships at sea will sometimes glow from electrical discharge from the sky. Airplanes will sometimes partially short circuit the atmospheric field and cause electrical discharges through the plane. In a different but related phenomenon, walking around the room on a dry day will cause the body to accumulate an electric charge, and touching metal or a ground rapidly discharges the field, and then we experience a spark.

ELECTRIC FIELD OF THE EARTH

There is an electric field from the earth to the sky. We walk through it all day. It varies with weather and surrounding objects.

LIGHT RAYS PASS THROUGH

Also flowing though the space before us are a large number of light waves. Every object in the room bounces some light back to all of us, and it all passes through the individual cubes which each of us has. For the sensor we use our eyes. Our eyes can pick up certain waves and focus them on the nerves in the eye. The nerves tell the brain and the brain recognizes the signal. If we point the cube in various directions, as "see:" various objects. It's all going on and radiating all the time from everything in the room. . This all means that at all times the space in front of your face is filled with multiple light paths, all crossing and intersecting one another, from every object in the room! It's all there and we are able to select and "see" some things, as we choose, and ignore others. We select which set of waves we will focus on or detect for the use of our brain. We do not see objects shadowed by walls or obstructions.

More About Light Waves

Light waves behave as undulations or fluctuations in space, and they are not molecular motions like sound. Light waves are short. The blue waves are shorter than yellow waves which are shorter than the red waves. Red or infra-red is the longest. The wave fluctuations come in tight bundles, and experiment and quantum theory show that in some types of measurement the waves behave like particles, called photons. The retina of our eye contains nerve endings, and these respond selectively to the various color wavelengths and photons, so that we are able to tell color. Wavelengths are measured in nanometers. Red is 700 nanometers, green is 400, blue is 300. Infrared is 1000 to 20000, and ultraviolet is 50-100 nanometers.

LIGHT RAYS PASS THROUGH

Everything in the room around us radiates or reflects light waves, of various colors. These pass through the cube. We "see" by selectively accepting and focusing some of these rays in the backs of our eyes.

RADIO WAVES PASS THROUGH

Radio waves are the general name given to electro-magnetic signals which are usually used to transmit intelligence from one point to another. There is a transmitting station and a receiving station, or receiver. Individual stations are assigned individual frequencies on which to operate. Station KCBS, for example, transmits at 740 KHZ, or 740,000 cycles per second. Radio waves can pass through things and through people. We detect these waves with the receiver. There are many radio waves form many stations going through the our cubic foot and through us all the time. With just a small radio we can hear broadcasts from all over the state, or around the country, and sometime from people around the world, just by listening in the proper manner. These signals are in my cube and also in your cube. Even if you put two human bodies around the radio, or two heads, the signals are still there and coming through us.

More About Radio Waves

Radio covers many frequencies and many types of signals. The lower bands of frequencies generally use amplitude modulation and are called AM. The AM band extends from 500 Kilo hertz t0 1.2 Megahertz. At various frequencies, usually higher, there are police radio, ship radio, aircraft radio, amateur radio. Other radio services, discussed later, include television, FM stations, and satellite stations. All these stations are transmitting simultaneously; and portion, or fragments, of their output can be found in the cube before your face.

RADIO WAVES PASS THROUGH

There are radio signals from all over the world passing through the cube. Place a radio receiver inside and turn it on and tune to different stations. People are talking into microphones and you hear them. Fields are there even when your hands or your heads are around the radio.

SHORT RADIO WAVES PASS THROUGH

This is another category of radio waves. These signals also go through our body. This category includes television, cellular phones, cordless phones, beepers, radar, and satellites over head. These systems are all busy putting out intelligent signals. If we have the right radio, or a TV set, we can pick them up, just from what is in front of us.

More About Short Radio Waves

Radio waves and light waves are made up of interacting rapidly changing electric and magnetic fields. The wave are called electro-magnetic waves or radiation. There is a continuous interchange of energy between the electric field and the magnetic field as the waves propagate through space. Radio waves travel at the speed of light. The various frequencies have different properties, some being able to pass through air and water, and some being attenuated or weakened. Antennas are designed to couple the electro-magnetic signals to space, or to the air. The antennas take different shapes for different frequencies. The higher frequencies have shorter wavelength and use smaller antennas. The highest frequency radio waves are called microwaves. Small antennas are workable, and much data can be sent over a small percent of the frequency. Transmission is quiet because there is not much natural or made-made "noise", or activity at these frequencies. A common source of microwaves are those used to prepare food. The microwaves penetrate the interior of the food and cause internal heating more rapidly than simply exposing the food to the surrounding heat of a oven.

SHORT RADIO WAVES PASS THROUGH

We sense these signals with a different radio. These include FM, cordless phones, satellites, cellular telephones, beepers, Radar.

LOCATION SIGNALS FROM SPACE

Other human-made transmitters send out electro-magnetic signals which tell us where we are on the earth. There are stations along the ocean coast line which transmit position or Loran signals to ships. Overhead. there are 15 special satellites that send out microwave navigation signals, right into a receiver held in our hand before our face. The signals tell us where a satellite is, such as over San Francisco or over Chicago or over New York City, and the receiver measures how long it takes for the signal to get here, and calculates exactly where we are in latitude and longitude. This system is the GPS system, or Global Positioning System. Ships and aircraft use it, and explorers all over the world use it. Some automobiles use it.

Some Typical City Locations Are

San Francisco: .... North 37 degrees 46 minutes; West 122 degrees 25 minutes.

Chicago: .... North 41 degrees 51 minutes; West 87 degrees 39 minutes.

New York City: .... North 40 degrees 43 minutes; West 74 degrees 01 minutes.

Houston: .... North 29 degrees 45 minutes; West 95 degrees 21 minutes.

Los Gatos
Unitarian Church: .... North 35 degrees 13 minutes; West 122 degrees 00 minutes

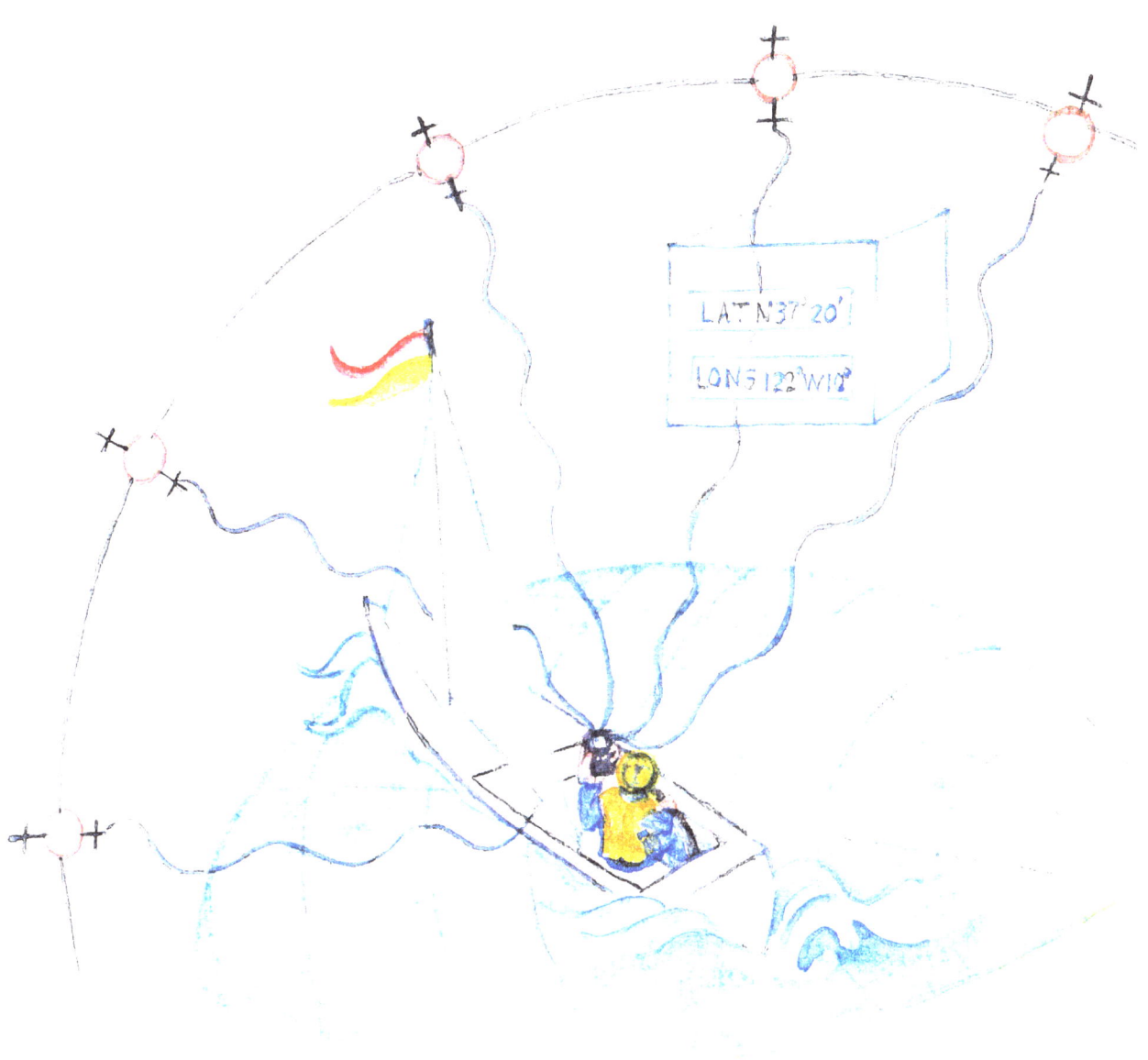

LOCATION SIGNALS FROM SPACE

There are special satellites in orbit above us which radiate radio location signals down to earth. These tell us where we are on the earth, in Latitude and Longitude. A Global Positioning System, called "GPS", instrument placed inside the cube will tell us where we are on the face of the earth..

HEAT WAVES PASS THROUGH

Heat makes it presence known in several ways. In hot air, the molecules move more quickly than in cool air. In a hot body, the molecules move more quickly. Heat is principally defined, however, by its radiation as an electro-magnetic field. Heat is transmitted from one side of the cube to the other. We can sense temperature with a thermometer or with our skin. If we light a match on one side of the cube, our face and our hand on the other side can tell that it is warm—in fact, it tells us not to get too close or we will be burned.

More About Heat

The wavelength of radiated heat is measured in millimeters. The electro-magnetic radiation frequency is higher at higher temperatures. In the electro magnetic spectrum the heat portion is referred to as the near and far infra red, meaning that it is just outside the red end of the visible spectrum. Hot bodies radiate a collection of frequencies in the infra red region. The infra red region overlaps with the highest frequency microwaves. Infra red frequencies, or the heat frequencies, will penetrate clouds, so that instruments sensitive to these frequencies can "see" some times better than visible light. The human body can "see" in the dark by sensing on our facial skin the presence of another warm human body.

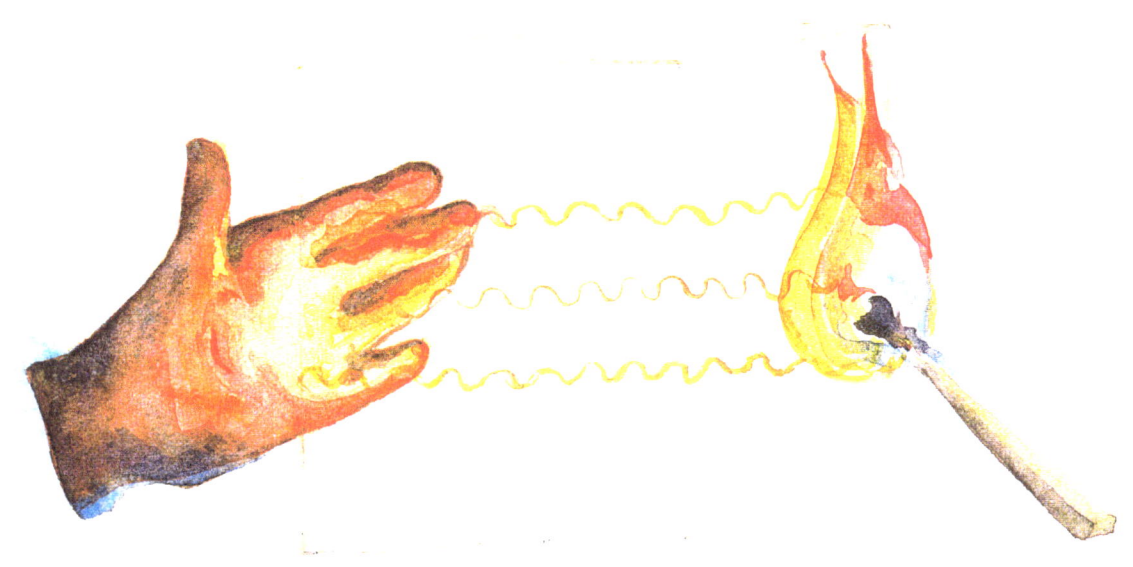

HEAT WAVES PASS THROUGH

Heat from hot objects passes through. We can sense heat waves with our skin.

X-RAY RADIATION AND COSMIC RAYS

There are X-rays going through our cubic foot. These come from stars millions of miles away in remote galaxies, or they may come from an X-Ray machine in a physician's or dentist's office. X-Rays penetrate our flesh but not our bone, thus making it possible to silhouette our bones on a film, and allowing study of broken bones. X-Rays penetrate cardboard and can spoil camera film. That's one reason film is wrapped in foil or metal, or a form of plastic which is impregnated to resist radiation penetration.

Cosmic Rays

There are cosmic rays going through our cubic foot. These come from outer space and from stars and galaxies millions of miles away. These can even shake up our body with atomic reactions, and cause oxidation of our bodies, and so some people conjecture that these rays make us grow old. We detect cosmic rays with an instrument called a Geiger-Mueller counter, which "fires" or goes "tick" when a cosmic ray strikes and ionizes atoms inside the tube.

X-RAY RADIATION AND COSMIC RAYS

X-Rays and cosmic waves are high energy radiation, generated by stars in outer space. These waves can penetrate the atmosphere, and human flesh. We are able to produce X-Rays ourselves

NEUTRINOS PASS THROUGH

Neutrinos come from the sun and penetrate our space. These are small zero-charge "lots of nothing" particles, generally the size of an electron. Neutrinos are produced within the sun as part of its radio-activity energy generation. 10,000 neutrinos pass through each square centimeter of our body every second. Most of these neutrinos don't hit anything or do any harm, and pass totally through the earth. There is a flux of neutrinos even in the cubic foot of space at the compressed center of the earth.

More About Neutrinos

The detectors for these neutrinos are extensive underground chambers using carbon tetrachloride (cleaning solvent) as a target. If a nucleus of carbon tetrachloride is struck by a neutrino, a secondary set of particles is generated, and these are detectable. Bursts of neutrinos from remote star formation have been detected. Not all scientists are convinced neutrinos actually exist. There is speculation that neutrinos may account for a large part of the mass of the universe.

NEUTRINOS PASS THROUGH

Neutrinos are particles generally the size of an electron and have no charge. Neutrinos mostly pass completely through the earth without collisions. Most neutrinos are generated by the sun.

FLAVORED MOLECULES

Our cube has other molecules and things besides air. Some can be detected using your nose as an instrument. You can mark your cubic foot with a perfume spray. Inject some spray. Each droplet contains several millions of molecules. Smell. The sense of smell has nerve receptors which can identify a great variety of molecules. The sense of smell can produce pleasure, such as from perfume or from fresh air, or can deliver warning signs of pollution, bad air, and poisonous gas. After your cube is marked with perfume, take it to another location and see if you still notice it.

More About Flavored Molecules

There are chemical receptors in the nose, in the epithelium, with nerve connections, which analyze and signal to the brain the identification of ingredients in the air. This is our olfactory sense. It is extremely sensitive. With concentrations of only 1 part per billion, we can detect differences in chemical composition, and molecules of various weights.

The sense of smell allows us to locate food, and locate mates, and sometimes can detect various emotions such as fear, or anger, or friendship, in the people we deal with. Animals and insects attract mates with suitable chemical airborne markers, known as pheromones. Each individual has small differences in chemical odor, and animal mothers use these differences to separate their own offspring from others. Trees emit airborne chemicals when attacked by insects and these emissions can signal other trees to raise defenses. Within your cubic foot, *you can even tell what the neighbors are having for dinner*!

The sense of smell is the bodies ability to use chemoreception from remote, usually airborne, sources. The sense of taste is chemoreception from direct contact. Taste is detection of chemical differences in food by the taste buds on the tongue..

FLAVORED MOLECULES

You can mark your own cubic foot. Flavor it. Add a few drops of perfume, using a spray bottle. Your nose is the detector.

PARTICLES AND LIFE IN THE SPACE BEFORE US

There are many molecules other than the oxygen, nitrogen, and carbon dioxide which are the basic constituents of air. There are trace molecules of argon, helium, much water vapor, and carbon monoxide There are particles barely visible. Hold the cube up to the sunshine—what do you see? Little bits and particles of dust floating around. Dust contains all sorts of debris. Dead skin, dead hair, and rubber particles that wore off from an automobile tire or asbestos from a brake. And on those particles, or flying free, there are small living creatures—bacteria, germs, fungus, spores, viruses, coming from animals, or from vegetation, or from humans. This life is looking for a spot to settle down and grow more life like itself.

More On Particles And Life In The Space Before Us

How do we sense bacteria? What ways can we sense these little animals? One way-let some one with a cold sneeze into the space, and ten people breathe it—someone will probably pick up the germs—and have a cold two days later. They may not know where they got it. A second way to detect small life: put some bacteria nourishment, on a dish, in the space before us. Some of these little animals will find it, and grow, and next day or two you will have green old or other color—something will be there that wasn't before. It will turn other colors, maybe green. The dish of bacteria food is called by scientists a Petri dish. A third way to detect microscopic life is to use a microscope... For particles and unusual molecules, use chemical analysis. With adequate equipment and knowledge, analysis of the atmosphere can reveal: what industries are present in the locale, what agriculture, what people have passed by, the time of year when the air specimen was taken, or what herbs grow in your garden.

PARTICLES IN LIFE IN THE SPACE BEFORE US

In addition to a variety of molecules before us, there are other particles, some alive, some dead, some ready to reproduce, some inanimate. Some particles carry bacteria. There are dust particles, which we can see in sunlight.

THE UNIVERSE SPEEDS THROUGH OUR SPACE

There is a background radiation of the entire universe. This radiation is in our cube, whether or not air and other radiation are present. The radiation is formed of electro-magnetic undulations and can be detected with high frequency radio receivers. We can tell by small frequency shifts that we are moving through this radiation field of the universe. The solar system goes through the galaxy at about 200 miles per second. The earth rotates, the earth travels around the sun, the sun passes through the galaxy, and then the galaxy goes through the universe. So the universe is whizzing by through this cubic foot of space, in various directions, all the time..

More On How The Universe Speeds Through Our Space

Scientists have discovered a cosmic microwave background radiation. The radiation is left over from the creation of the Universe in a Big Bang 15 billion years ago. The radiation is detected with a microwave radio receiver. The antenna of that receiver can be placed inside your cube and the values measured... Space speeds by or races through your cubic foot—or does our cubic foot speed through the universe? The answer is that science always chooses the most simple answer, and it is more simple to assume that your cube moves, while the main part of space or the universe stands still... The universe has a temperature corresponding to the background radiation frequency of 300 GHZ. It is cold, but not zero. The average wavelength is 1 millimeter...How fast do we go through the universe? Well—the earth rotates once every 24 hours, and we go 25,000 miles with it, so that is about 1000 miles per hour due to rotation. Then the earth goes around the sun, and that is 20 miles per second. Then the entire solar system goes through space at 200 miles per second. I hope we don't hit anything. Who is driving? Well, sometimes we do hit rocks and stuff in space, and then we see shooting stars...

THE UNIVERSE SPEEDS THROUGH OUR SPACE

The Universe has a temperature. We can measure the temperature by its radiation and we can tell we are moving through this radiation. As our galaxy moves through the universe this radiation can tell us how fast we are moving.

THINGS NOT FOUND IN OUR SPACE

Human kind has conjectured and imagined and dreamt of other creatures about us. What about spirits, nymphs, ghosts, ectoplasm, goblins, devils, flying carpets, leprechauns, zodiac creatures, even other dimensions, and the like? We sometimes think we "feel" or sense such personalities. No one has been able to establish their existence, or confirm that what one person "feels" is the same as another feels. The feeling is not repeated whenever one wishes to make an experiment. Tangible things in the space around us can be found and measured over and over again by other people and found to behave in exactly the same way. in this tangibility or measurability we can have "faith". Most thinkers, therefore, believe that for all practical purposes, these immaterial creatures exist only in our imagination, from wishful thinking, and not in the universe. So far all searches have been a waste of time.

More About Things Not Found In Our Space

Think that just possibly, *after all the other* puzzles and behavior of the physical universe have been worked out, and all the puzzles of the human mind and body worked out, we just might find deeper mystery and magic and spiritual beauty remaining to be studied. For now, we must do our basic work.

THINGS NOT FOUND IN OUR SPACE

No one has found a detector for spirits, nymphs, ghosts, ectoplasm, goblins, devils, angels, flying carpets, leprechauns, zodiac creatures, and the like. There's nothing we can measure there.

Black Hole

EXTRA SENSORY PERCEPTION

We can't measure Extra Sensory Perception. It would be exciting to find it. Maybe there is extra sensory perception—maybe we can read other minds by signals we don't understand yet. People report such events. Some thing's happen in the space which we aren't sure of. Sometimes people see things in a dream and the next morning they wake up and find these things really did happen. How did the message get there? Not by anything we know in other ways. We are unable to confirm or repeat the action.

More About Extra Sensory Perception

Maybe we have what is called extra sensory perception. Our other sensors and detectors don't help us. No one has a detector that works all the time—only once in a while. Our sensors, if any, are intermittent. We don't have the instruments to measure it—or we don't know how to turn on the sensors we have—or there isn't any thing there. All the reports might be merely wishful thinking, or illusions, or hallucinations, or chicanery, or coincidence. People have tried to find ESP using marked cards, or seances, or trying mentally to control how dice roll. So far, no dice.

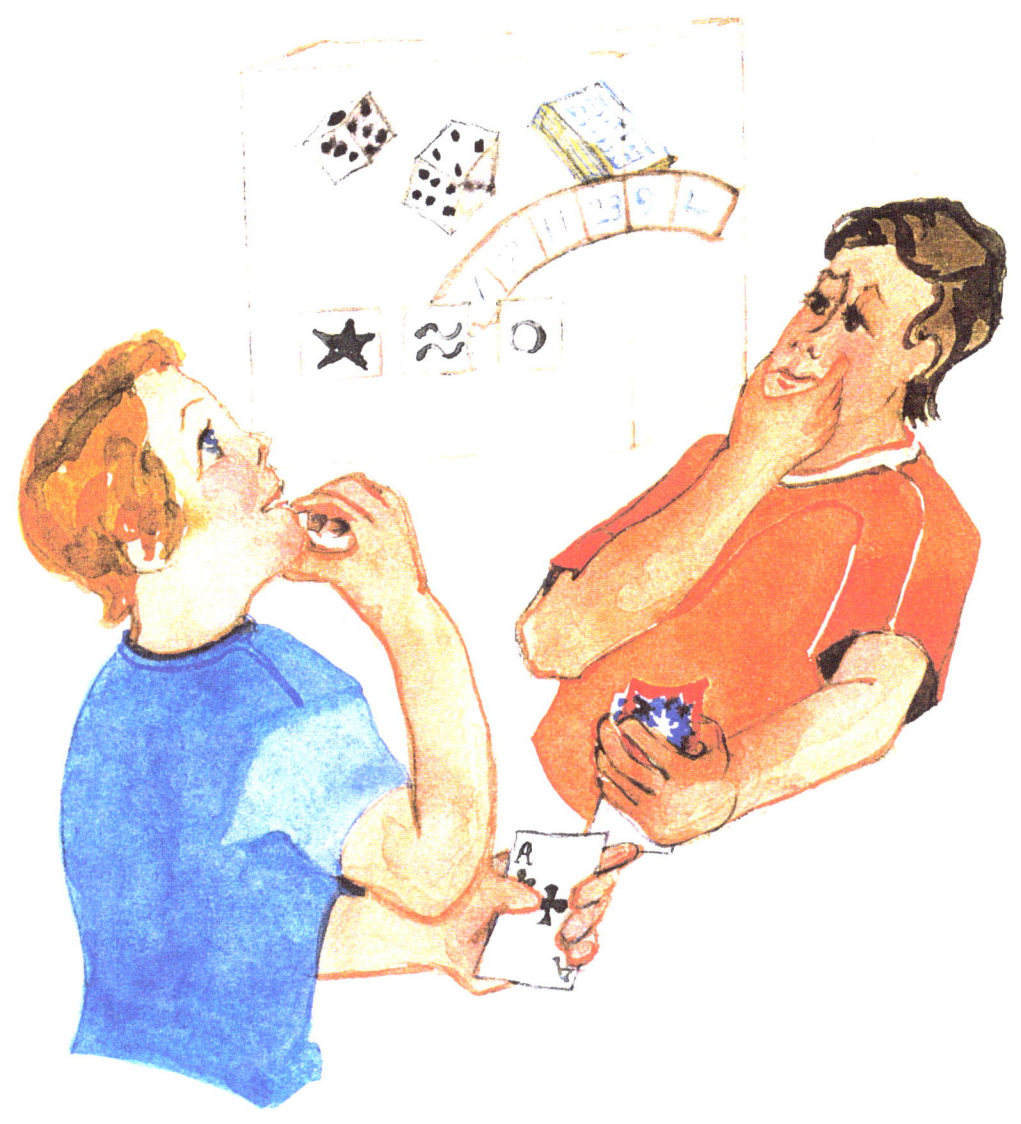

EXTRA SENSORY PERCEPTION

Extra Sensory Perception, or ESP, is sensing activity beyond our usual five senses, in ways not understood. We suspect it might be possible to communicate in mysterious ways, or know the future, without our regular senses. This is exciting but unproven—and for the time being at least, upriver.

MIND TOUCHES MIND

Last—what else goes through our cube? Point your cube at anyone in the room. Let your thoughts reach out to a face. Somehow behind the face your mind touches that other mind. Try moving your cube to—a friend, a parent. or someone you know. Do you sense a tingle? Something good happened. Your mind responded and sent chemical messages to your body. Your mind makes available a host of information about that other person, whether they are well known to you or freshly met. We can communicate love and friendship whenever we meet someone, right through the space in front of our face! Some of that feeling happens between humans and animals. Is it just light rays and sound waves, and the nerves in our eyes and ears, or is there something more, something subtle but important, waiting to be discovered? When we find life on other planets, will there be an intuitive response?

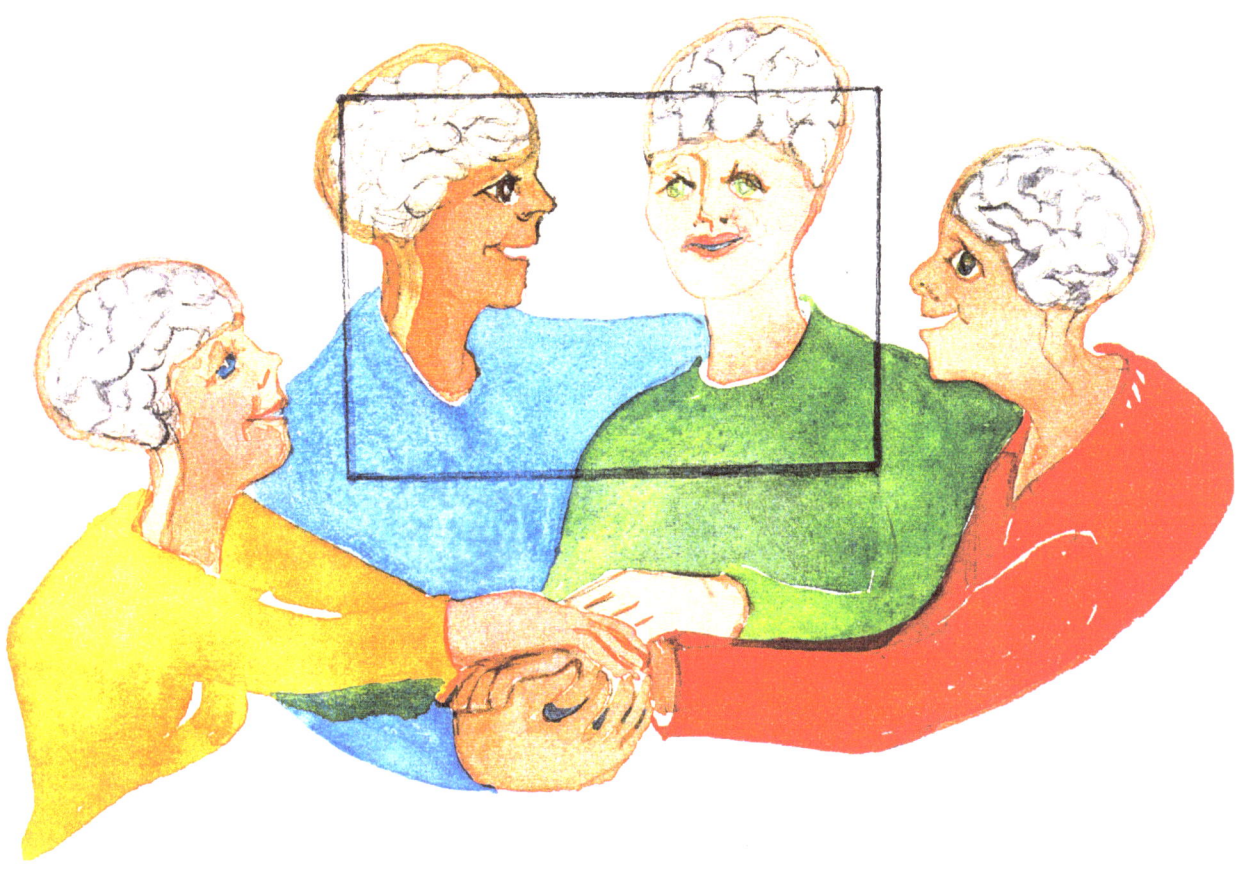

MIND TOUCHES MIND

Our mind reaches out to another person. We see another face and behind that is a mind. Our minds touch. We see a touch of a smile—they see ours—we sense another mind, we tingle a little bit, and we sense the joy of knowing other human beings.

A LOOK BACK

A great deal goes on in the "empty" space before us. Twenty phenomena have been described and there are more. Different instruments detect different phenomena. All the instruments convert otherwise invisible phenomena to forms readily observed with our senses, usually sight and sound.

Some phenomena are detected by gathering a large number of samples and observing the bulk effect. A telescope gathers a large number of faint signals and focuses them on a spot on a photographic plate. A biological effect is detected when a large number of organisms grow in a culture (Petri) dish. DNA samples are detected when a large number of copies are grown in test tubes, and then separated on to a visible strip made visible by radioactive means. The various electro-magnetic phenomena are usually detected by setting up a resonant system which responds to a large numbers of repeated cycles. Each radio station or each TV channel has a carrier frequency which always has the same repetition rate. Large numbers of cycles with the same time spacing, or same frequency, are sent out, and each station has its own allocated frequency to distinguish it from all others.

Other phenomena are steady in nature, or "DC" (for direct current). These include gravity fields and magnetic fields. Conversely, some phenomena are studied by observing the traces of individual particles—these are usually atomic or nuclear reactions. For these short lived phenomena, scientists rely on repeating the experiment several times before being certain of the results.

But within the space before us, all the laws of physics apply, as also do the related arts of chemistry and mathematics. We learn much about the universe by studying our immediate vicinity, or letting the phenomena report to us. We learn much about the mind, and about biology, by studying ourselves. But trips to the north and south poles, to the bottom of the sea, to the moon and beyond, are important in order to get a full understanding of some phenomenon—and in order to fulfill humankind's destiny.

A LOOK BACK

Much goes on before us. It is great entertainment. The more we know the more amazing it becomes!

NOTES

MEASURING THE SPEED OF LIGHT

Fast People Turning Off the Light

A famous basketball player was noted for his great speed and lightning reflexes. They asked him how fast he was. "I am so fast that I can switch off the light and be in bed before the room goes dark." What is the interval of time from removing electric power to the bulb and darkness in the room? Assume the bulb cools instantaneously. Assume the distance from the bulb to bed is 20 feet. Calculate what the interval of time is for light (or darkness) to cover the 20 feet (6 meters)

Light travels at 186,000 miles second, or 300×10^6 meters per second. This speed turns out to be one foot per nanosecond. More precisely, light travels at 298.853×10^6 meters/sec. and .98 feet in one nano second.... If the basketball player travels 20 feet to get into bed, he has to do the trip in 20 nanoseconds, just to be there as the light dies. That is .02 microseconds. Not much chance. And, further, Einstein's law (no longer a theory) show that he would need to achieve nearly infinite mass to reach the speed of light, so it doesn't look likely the feat could be accomplished.

Early Attempts at Measuring the Speed of Light

Early scientists studied the speed of light. Suppose we station a person on a hill top with a lantern and a light shield and a stop watch. Station an assistant a mile away on top of another mountain or a tall building, also with a lantern and a light shield. The first person unshields his/her light, and starts his top watch. The remote assistant upon seeing the light unshields his light. The first person records the time of the return signal. Now we know the time for a round trip of the light beam.

What are the results of this experiment? One finds that the calculated speed is the same whether the assistant is one mile away or a 100 feet away. The reason is that the human reflex time of exposing a light is far larger than the travel time for the light beam. The human time might be, for example, ½ second. The light travel time is only 10 microseconds for the two mile round trip. . The measured round trip time of 500 milliseconds is 50,000 times longer than the time for the light trip speed, and so the experiment has only measured human reaction time, and the light portion is missed.

Light from a Jupiter Moon Traverses the Diameter of the Earth's Orbit

More sophisticated methods of measurement are needed. To measure high speeds a long measuring line or base line helps. Such a line is the diameter of the earth's orbit around the sun. The following procedure was suggested by a student of Galileo's. Galileo no doubt surrounded himself with extra smart students, or smart students find the professor. The person doing the beset and earliest work was Ole' Romer working in the Paris Observatory in 1676.

A source of blinking light is the rotation of the moons of Jupiter around Jupiter. The moons appear and disappears and reappear, in a known way whose time can be established. Further, the earth to Jupiter distance varies during the year, by a known number of miles, as the earth circles the sun, and as our seasons progress. So the blinking light from Jupiter is sometimes nearer and sometimes further away. Measure this time difference and now we can compute the time for the light to traverse the diameter of the Earth's orbit. Seek out this time difference and we can compute the speed of light.

See the drawing in Figure 1. Pictured is the moon of Jupiter circling Jupiter, appearing and reappearing. At the receiving end of this blinking light is the Earth, circling the sun, having maximum and minimum distances from Jupiter. The radius of the earth's orbit is 93 million miles, and the diameter is twice that, or 186 million miles

Needless to say, such a procedure is far more easily said than done. The observer spends countless nights studying the appearance of the moons of Jupiter, then comparing this time with a reference time, then estimating the orbit of the earth, then adjusting for the position of Jupiter. A problem is that the time differences due to light travel are measured in minutes, and yet the time readings to be compared are six months apart, as the earth goes around the sun. The best pendulum clocks are not accurate enough:.. even the best pendulum clocks lose or gain in seconds per day. But there is one large inertial object that rotates steadily with no visible perturbations, and that is the Earth. The Earth's rotation serves as a clock. One can study sunset and sunrise times, and meridian times. That too is easier said than done, since sunrise and sunset times vary through the year. The peak at midday of the sun or of a star at midnight is not so easy because it requires an accurate vertical reference and because there are necessary latitude adjustments.

The period of the selected moon of Jupiter is not an integral function of the Earth's rotation. Multiple readings and careful records are necessary. The method was used, however, and yielded a value with an error of less than 20%, which isn't too bad considering the problems. For the speed of light, getting the decimal point right is an achievement.

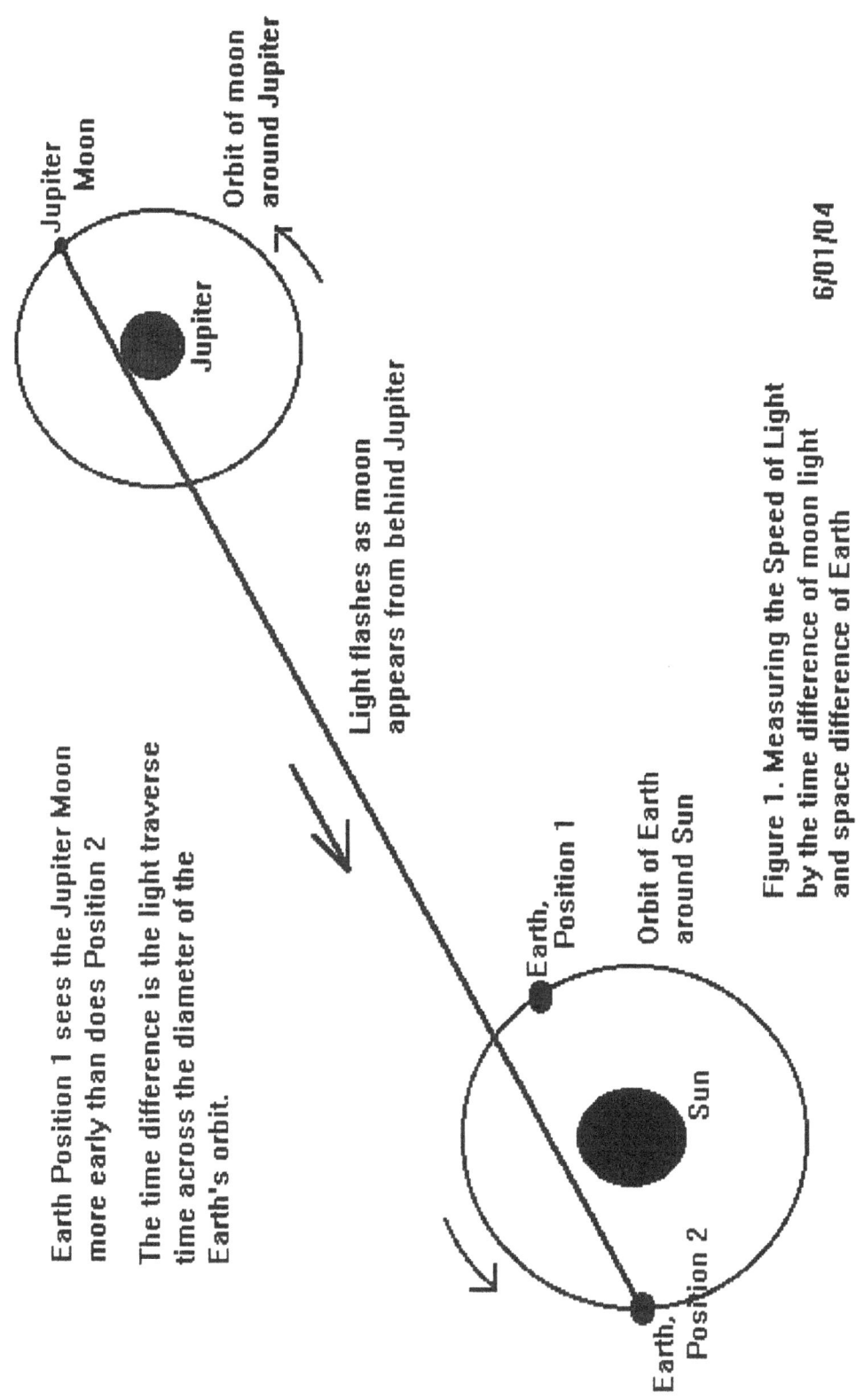

Figure 1. Measuring the Speed of Light by the time difference of moon light and space difference of Earth

Vibration of Electric and Magnetic Fields

Another value for the speed of light was obtained by James Clerk Maxwell, a Scottish physicist, using an unexpected method. Light is a vibration, analogous to the vibration of air molecules from sound, or of a plucked guitar string. The equations of vibration for air molecules or for a guitar string are known, and are based on such factors as the tension, length, and the inertia of each segment of the string. The vibrations travel back and forth along the length of the guitar string. The heavier the string, the more slowly it vibrates, and the musical tone is lower. The tighter a string, the more rapidly it vibrates, the tone is higher. This happens when a guitar is tuned.

A light beam is made of the interaction between electric fields and magnetic fields. An oscillating electric field creates an oscillating electric current, and the current forms at right angles an oscillating magnetic field. A moving magnetic field generates an electric field. Michael Faraday and Joseph Henry showed these interactions. Maxwell knew the arithmetic values relating these values, just from static experiments on a laboratory bench. He also knew the equations of vibration. Plugging these numbers together, again much easier said than done, he was able to find a background number which was equal to the velocity of light. His value also was within 20% of the correct speed.

A Blinking Light Driven by a Motor

Suppose we have a rotating disc with a slit in it, and a light behind the slit. As the disc rotates, a flash of light is emitted at each revolution. This light is sent to a remote mirror, and reflects back to the disc, a short time later. The mirror reflection is virtually instantaneous, compared with the lantern method described in the first system. We look for this reflection through the same slit on the rotating disc, and we move the telescope in the direction of rotation until the maximum response point is found. We notice that the slit has moved a bit during the round trip transit time of the light flash. The time is found from knowing the rotation speed of the disc and the number of angular degrees of motion between send and receive flashes. We can measure microseconds this way, and we can measure how far away the mirror is, and so we can find the speed of light. It was Albert Michelson, a young man in the Navy, who got credit for the first accurate figures for these measurement. The mechanism is sketched on page 47.

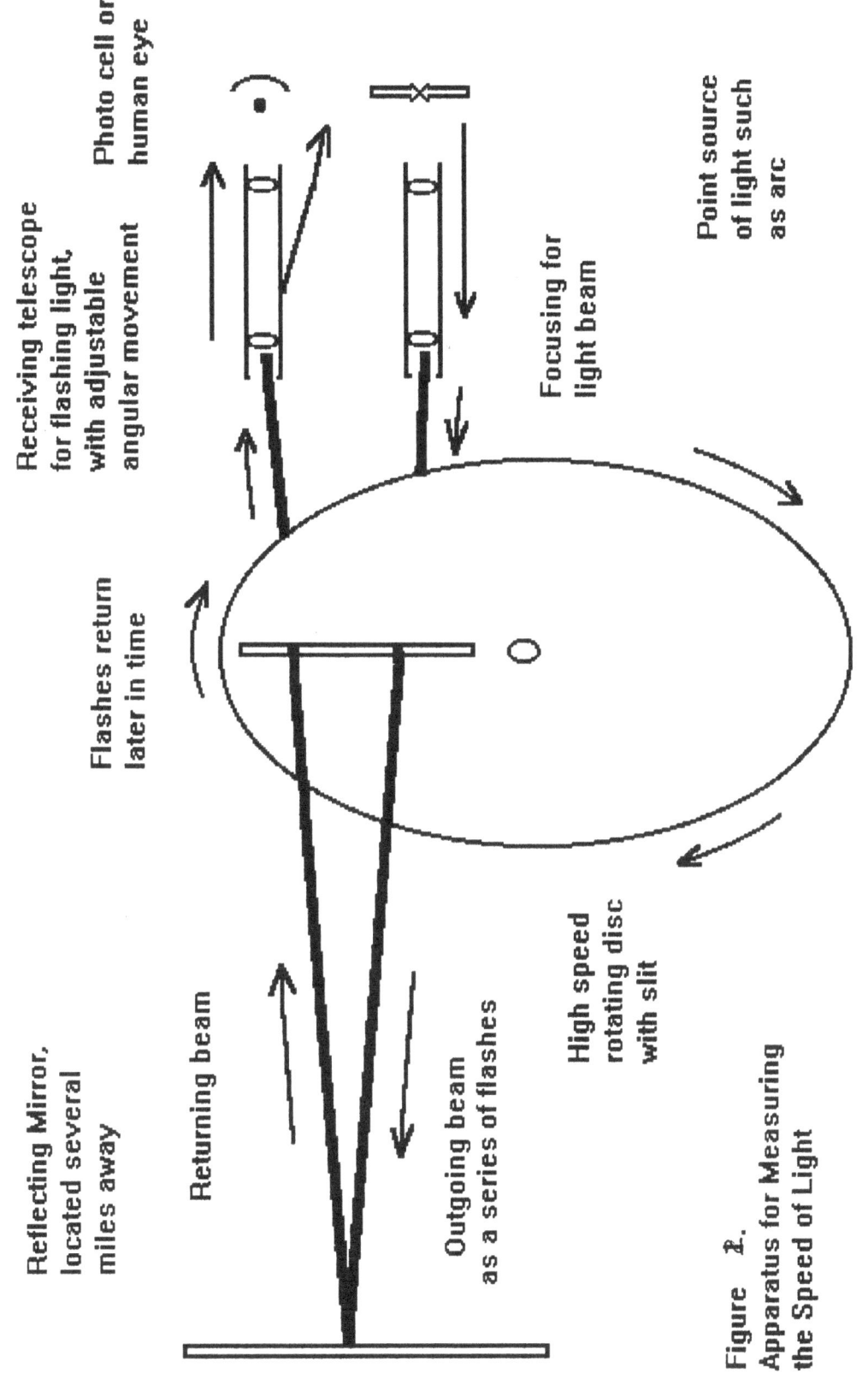

Figure 2.
Apparatus for Measuring the Speed of Light

Measurement of the Speed of Light with a Conventional Oscilloscope.

The oscilloscope found in many electronics labs today can measure the speed of light. Oscilloscopes generate a time base. Flash a light on a photocell to starts the time base moving. Point a telescope at a remote mirror to see the remote reflection. The telescope connects to a second photocell, which makes an electric pulse to show on the oscilloscope screen. The location of the pulse tells us how long it took for the round trip of the light beam. Now we know the speed. I felt proud when I did it. Bouncing electro-magnetic waves similar to light is what makes radar work.

wouldn't run just as well in such conditions. A clock follows the laws of physics and its inertial timing mechanism follows its own regular space-time continuum. We don't have a picture of time to show you, yet we know that inside our cube it does.

CUBES AND DIMENSIONS

We ordinarily know three dimensions. Our cube has three dimensions, in which things are placed, and in which we can move around. A dimension has a direction and it can be positive or negative. One direction is left and right. Another direction is up and down. A third direction is forward and backward. These three directions are perpendicular to one another. Each dimension has a direction which does not infringe on the other dimensions.

There are three lines radiating from any corner of our cube. If you look at the lower left hand corner of the cube, then you see a line to the right, a line up, and a line going backwards. These are referred to mathematically as the **X** axis, or **X** dimension, the **Y** axis, and the **Z** axis.

What about a fourth dimension? How can we understand it? One point of view says it is time. Do this thought experiment, which is an experiment in your head. Imagine a solid object, such as an apple. Slice through it with a sharp knife, or a plane. What is the result? A surface, or a two dimensional representation of part of the apple. Next imagine slicing the plane with a second plane. The result is a line, or a single dimensional representation of part of the apple. Next slice through a line. The result is a point. What happens is that each slicing action produces an example of a structure with one less dimension than before.

TIME IS SUGGESTED

Following this line of thought, what could be sliced through to produce three dimensions, or "now"? "Now" has three dimensions. Well, it might be time. Slice a piece of time, and we have a still photograph of us. Slice through time at another instance, and we have a different still photograph. So therefore the fourth dimension is time? Quite likely. Much of modern physics deals with describing the "space-time continuum" of a particle or an object or an event. "Space –time" combines the **XYZ** dimensions with time, **T**.

Time is different from the other three dimensions. In the **X**, **Y**, and **Z** dimensions, we can move back and forth under our control, but with time we have no choice but to go forward, propelled beyond our control.

AN EMPTY (?) CUBE

In this book, we show what goes on in the cube of space before us. Suppose we put the cube where these pictured things are not going on. Put the cube in remote intergalactic space, billions of miles from any galaxy, where the gravitational fields are remote and weak and self canceling (there are some special spots that are totally self canceling). Put it in a metal shell to shield out all electromagnetic effects. Put it in a high vacuum to keep out all particles. What still goes on in the cube? Well, time still marches on. No one has checked this experimentally, but there is no reason to think a clock wouldn't run just as well in such conditions. A clock follows the laws of physics and its inertial timing mechanism follows its own regular space-time continuum. We don't have a picture of time to show you, yet we know that inside our cubs it does.

HYPERCUBE

There is another path to understanding dimensions beyond that of the cube. In the following analysis, the fourth dimension is geometric. What does it look like? Is it possible for us even to imagine an extra physical dimension? It doesn't seem possible, but a description is possible, and mathematics points the way.

Write a simple expression: **L+2**. L represents a linear dimension or a line. The "2" represents the two ends of a line. We now calculate the effect of raising this expression to various powers, representing progressively increasing dimensions.

First Step: In the zero-th dimension, using the symbol "^" to represent an exponent, and the exponent is zero, **L+2)^0 equals 1**, meaning that one dimension-less point represents this initial dimension.

Second Step: Use the exponent "1" to define a one dimensional object.

The value of **(L+2)^1** is **L+2**. Interpreting, the first dimensional object has a single line L with two terminating ends or points.

Third Step: Use the exponent "2" to define a plane, or a two dimensional object.

(L+2)^2 equals: **L^2 + 4L + 4**. The significance of this result is that a plane, **L^2**, has 4 sides L and 4 points or vertices. These steps show that the formula is predicting correctly. Look at **Appendix 2** to see how the algebraic calculation is done.

Fourth Step: Use the exponent "3" to define a cube, or a three dimensional object.

(L+2)^3 equals: **L^3 + 6L^2 + 12L + 8**. A cube **L^3** has 6 planes **(L^2)**, 12 lines L or edges, and 8 vertices, or points. Check it. Pick up a cube and count its qualities.

Fifth step: Use the exponent "4" to define our hypercube, or a four dimensional object.

L+2)^4 equals: **L^4 + 8L^3 + 24L^2 + 32L + 16**. Therefore, we know that a four dimensional hypercube has: 1 hypercube **L^4**, 8 cubes **L^3**, 24 planes **L^2**. 32 lines L, and 16 vertices or points.

What does such a geometrical figure "look like"? Well, have a shot at it, but generally it is beyond our limited minds to visualize. Some other-worldly creatures may be able to do it.

Similarly we can predict higher order dimensional "appearances". If you like algebra, take the derivation in the appendix one step further.

The hypercube is also known as a "*tesseract*" and it is also known as a fourth dimensional cube.

ALGEBRAIC EXPANSION OF DIMENSION FORMULA

To do algebraic multiplication, the steps are similar to numerical "long" multiplication, with the exception that alphabetical identification is preserved.

To square $\underline{L+2}$, multiply $\underline{L+2}$ by $\underline{L+2}$, as below:

$$L + 2$$
$$L + 2 \quad \textit{Multiply by } \underline{L+2}$$
$$\overline{L^2 + 2L}$$
$$ 2L + 4$$
$$\overline{L^2 + 4L + 4} \quad \textit{Defines a Square}$$

To cube $\underline{L+2}$

$$L + 2 \quad \textit{Multiply by } \underline{L+2}$$
$$L^3 + 4L^2 + 4L$$
$$ 2L^2 + 8L + 8$$
$$L^3 + 6L^2 + 12L + 8 \quad \textit{Defines a Cube}$$

To get the 4th Power of $\underline{L+2}$

$$L + 2 \quad \textit{Multiply by } \underline{L+2}$$
$$L^4 + 6L^3 + 12L^2 + 8L$$
$$ 2L^3 + 12L^2 + 24L + 16$$
$$L^4 + 8L^3 + 24L^2 + 32L + 16 \quad \textit{Defines a hypercube}$$

GLOSSARY

Algebra .... a branch of mathematics in which letters are used to represent numbers, quantities, and mathematical entities.

Background

Radiation .... the mixed radiation which exists other than the specific radiation being studies.

Bacteria .... Single celled prokaryote organism, lacking a nucleus and organelles; contrasted with eukaryotic organism, known as protista, which are single celled with a nucleus.

Cosmic .... the universe, in contrast to the earth alone, or vast extraterritoriality

Cosmic Ray .... atomic nuclei of extreme penetrating power and with energy levels of one billion to a million billion electron volts.

Cube .... the regular solid of six equal square sides.

Ectoplasm .... the emanation from a spiritualistic medium and that is believed to affect telekinesis and similar phenomena.

**Electromagnetic
Waves** .... waves that are propagated by simultaneous periodic interaction of magnetic and electric fields, the two fields being perpendicular to one another, and having a propagation speed of that of light, 300×10^6 meters per second.

Extra Sensory .... beyond the ordinary or known senses, such as mental telepathy.

Fourth Dimension .... a dimension beyond the three ordinary dimensions of length, width, and depth, and which in relativity theory is time.

Gamma Ray .... a photon or radiation quanta emitted spontaneously by radioactive material, with wavelength from 1.0 to .01 picometers.

Gravity .... an attraction between two bodies that is proportional to the product of their masses and inversely proportional to the square of their distance, resulting for free bodies of an acceleration towards one another.

Infra-red .... electromagnetic radiation similar to light lying outside the visible spectrum and of wave length 1,000 to 30,000 nanometers, effective in heating bodies on which it falls.

Leprechaun .... a mischievous elf of Irish folklore, believed to have knowledge of hidden gold.

Light Waves .... electromagnetic waves of the visible spectrum, generally 380 (violet) to 770 nanometers (red) in length.

Magnetic Field .... a field generated by macroscopic or atomic electric currents, in which both magnets and electric currents experience a mechanical force, and which can generate electric current when felt by a moving electric conductor.

Molecule .... a unit of matter that is the smallest particle capable of retaining it's identity with the substance in mass.

Neutrino .... an uncharged elementary particle, equivalent to the electron, interacts weakly with matter, and has one-half quantum of spin. Neutrinos have bad breath.

Space .... needed to hold matter and energy, and believed to be constantly expanding; also the abode of the mythical kilroy.

Time .... nature's way of preventing everything from happening all at once.

Unicorn .... a mythical animal with a single horn at the center of the forehead, the body of a horse, and the tail of a lion.

X-Rays .... electromagnetic radiation, similar to light, of wave length from .04 to 10.0 nanometers, highly penetrating.

APPENDIX

ALGEBRAIC EXPANSION OF DIMENSION FORMULA

To do algebraic multiplication, the steps are similar to numerical "long" multiplication, with the exception that alphabetical identification is preserved.

To square $\underline{L+2}$, multiply $\underline{L+2}$ by $\underline{L+2}$, as below:

$$L + 2$$
$$\underline{L + 2} \quad \textit{Multiply by } \underline{L+2}$$
$$L^2 + 2L$$
$$\underline{2L + 4}$$
$$L^2 + 4L + 4 \quad \textit{Defines a Square}$$

To cube $\underline{L+2}$

$$L + 2 \quad \textit{Multiply by } \underline{L+2}$$
$$L^3 + 4L^2 + 4L$$
$$2L^2 + 8L + 8$$
$$L^3 + 6L^2 + 12L + 8 \quad \textit{Defines a Cube}$$

To get the 4th Power of $\underline{L+2}$

$$L + 2 \quad \textit{Multiply by } \underline{L+2}$$
$$L^4 + 6L^3 + 12L^2 + 8L$$
$$2L^3 + 12L^2 + 24L + 16$$
$$L^4 + 8L^3 + 24L^2 + 32L + 16 \quad \textit{Defines a hypercube}$$

APPARATUS LIST

Demonstration equipment is helpful while reading this book, such as:

1. Crumpled pieces of newspaper, to show air molecules flowing.
2. Barometer, to show air pressure.
3. A small apple, to show gravity.
4. Compass, to show earth's magnetic field.
5. Prism, to show some of the properties of light.
6. Small radio, to show radio stations, long and short wave.
7. TV receiver, to show TV signals.
8. GPS Receiver, to find latitude and longitude from satellites.
9. Perfume spray, to "mark" a cube of air, using the nose as an instrument.
10. Other available instruments are the eyes, ears, nose, and touch.
11. PVC corner members, three way, 1/2" size, quantity 8.
12. PVC members 1/2" diameter, each 12" long, quantity 12.

To be included with later editions: Compass, plastic apple, sphere of Styrofoam, small radio ($6.95 retail), perfume dispenser, plastic parts for forming a cube from drinking straws.

TO THE READERS

Your feedback is welcome and appreciated. There will be another edition, to incorporate the results of your comments.

Charles Walton
19115 Overlook Road
Los Gatos, CA 95030
Tel: *408) 354-1358 • Fax: 354-8917

IDEAS

NOTES

www.ingramcontent.com/pod-product-compliance
Lightning Source LLC
Chambersburg PA
CBHW051044180526
45172CB00002B/514